THAT'S A JOB?

I Like SPACE

... what jobs are there?

by Steve Martin

Illustrated by Tom Woolley

Kane Miller
A DIVISION OF EDC PUBLISHING

CONTENTS

Planetarium
educator 27

Space weather
forecaster 28

Public affairs
officer 30

Scientific
researcher 32

Science fiction
author 34

Technical
writer 35

Space suit
designer 36

Computer
engineer 37

Astronaut
instructor 38

Neuroscientist 39

Equipment
specialist 40

Technician 41

Project manager 42

Your perfect job match 44

There's more ... 46

INTRODUCTION

Qualities and skills you need to explore space

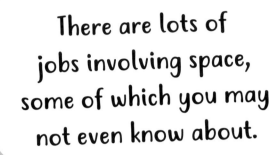

There are lots of jobs involving space, some of which you may not even know about.

Do you dream of blasting off into space? Are you fascinated by the idea of life on other planets?

Everyone knows that astronauts work in space, but you'd be amazed at how many other jobs there are for people with a passion for the stars!

Whatever your skills and interests are, there's sure to be a job to match. Astronauts need to be brave, astronautical engineers need to enjoy making and fixing things, science fiction writers need to be creative, and astronaut instructors need to love working with people.

But there are also qualities that are essential for everybody working in space: you must be motivated, curious, and love learning. Space is a very, very big place, and we still know very little about it—but that's what makes these jobs so exciting!

Many of the jobs also need good team players. It takes a lot of people to put an astronaut in space, and everyone has to play their part. Team players are reliable, responsible, willing to listen to others, and always ready to lend a hand.

There are lots of jobs for people who enjoy science, computers, or engineering, from astrophysicists and materials scientists to computer engineers and space suit designers. Not every job needs scientific skills, though. For example, space agencies employ people to tell the public about the work they do, and space lawyers use their knowledge of the law to advise people.

Whatever the job, if you are fascinated by our amazing universe and passionate about learning new things, there's a whole world of discoveries waiting for you!

If all this sounds like you, then you're the right type of person to work in space!

This book *counts down* 25 different jobs involving space. You'll learn all about what you need to do to make your dream job *take off*, giving you a sneak peek into a typical day in the life of each worker. You'll learn the important stuff, such as what it takes to get the job and what duties and tasks are involved, and you'll discover the fun stuff too, such as how to look for life on other planets, and the worst part of an astronaut's job ...

HINT: It involves going to the bathroom!

When you've read about all the different jobs in the book, turn to page 44 to find out which jobs might suit you, or page 46 to discover even more jobs!

ASTRONAUT (COMMANDER)

I still can't believe I was selected to become an astronaut. It took many years of study before I could even apply for the job, and thousands of other people applied. Now, I'm a commander. That means that I'm in charge of the Space Station—a crewed spacecraft that orbits (moves around) Earth. It's a lot of responsibility, and an incredible honor.

I earned a degree in engineering and then completed further study in astronautical engineering to become an astronaut. I'm also an experienced aircraft pilot. Even after I was selected, I had to do several years of training. But my hard work was worth it, and I'm now doing my dream job!

1

As the command astronaut on board, I'm in charge of the station and the five other astronauts here at the moment. My first task is to call Mission Control Center on Earth to discuss my work for the day. Later, we'll be going on a space walk around the outside of the Space Station to repair some equipment. It's my favorite part of the job!

2

Another astronaut will be coming with me on the space walk. We put on our space suits as we talk about what we'll be doing. We talk to each other in Russian, his country's language. People from all over the world work at the Space Station, so speaking other languages is very useful in my job.

3

Our suits are on and we're ready to enter the airlock. It has two hatches: we enter through the first hatch, making sure it is locked tight behind us to stop air escaping when we open the second hatch.

4

The second hatch opens ... and we're in space! We're attached to the Space Station by a rope called a tether so that we don't float away.

5

We get to work right away, making repairs and checking the equipment. We listen to instructions from Mission Control through the radios inside our helmets. We grab on to handles on the side of the Space Station to move ourselves around.

6

We stop for a moment to admire the view of Earth—it's incredible! I feel so lucky to be able to experience this. Finally, after six hours, our repairs are finished, and we reenter the airlock. We're exhausted but happy that the mission was a success!

7

I eat dinner with the crew, do some exercise, and then head to bed. The station orbits Earth at 17,000 miles per hour, and we see 16 sunrises and 16 sunsets every day, so our sleep times are carefully planned. I climb into my sleeping bag, attach it to the wall so I don't float off, and fall fast asleep ...

SPACE LABORATORY

The International Space Station is the astronauts' main workplace, where they do experiments and take care of the spacecraft. Although it's about 200 miles above us, it can be seen as a bright light moving through the sky. You have to know where to look though, since it circles Earth every 90 minutes.

MY JOB: BEST AND WORST PARTS

BEST: I'm proud to be an astronaut. It's the coolest job in the universe!

WORST: My space suit has a special diaper so I can use the bathroom. It takes a lot of getting used to!

ASTRONAUT (FLIGHT ENGINEER)

I'm a fully trained astronaut and a highly qualified scientist. I work on board the Space Station, alongside other astronauts from around the world. As a flight engineer, I'm responsible for lots of different tasks, such as carrying out scientific experiments and operating equipment.

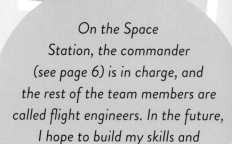

On the Space Station, the commander (see page 6) is in charge, and the rest of the team members are called flight engineers. In the future, I hope to build my skills and qualify as a commander.

1

I start the day by washing up—with a pouch! To save water on board, we use special soaps that don't need rinsing. After a quick breakfast, I talk to the team about what we'll be doing today.

2

We have just received some sample materials from our scientists on Earth, and my first task is to install them on the outside of the Space Station. We want to find out how certain materials, such as metals, plastics, and fabrics, are affected when exposed to outer space. This will help us make sure we're using the right materials to build and protect our spacecraft.

3

I get the case of samples ready and fix it onto a robotic arm. The robotic arm carries the case through the airlock hatch and out into space.

4

The new case is fastened into place and the samples are uncovered. The samples will stay out there for a year! The robotic arm removes another case of samples we put there six months ago. The samples will travel back to Earth on a space shuttle, to be tested by a materials scientist (see page 9).

5

The task is completed! Before our evening meal, I check on another experiment: testing how plants grow in space so we can have fresh fruits and vegetables on board. But for now, we eat freeze-dried vegetables for dinner!

MY JOB: BEST AND WORST PARTS

BEST: I love working in the universe's most exciting laboratory!

WORST: Being away from my family and friends for months at a time can be tough.

MATERIALS SCIENTIST

I've always loved science, especially doing experiments! In my job, I study different kinds of materials, such as metals, plastics, and fabrics. I find ways to improve them for use on spacecraft and for space suits, so that they last for a long, long time.

1

This morning, I'm very excited! I'm getting a new delivery that has been sent from the Space Station. It's a case of sample metals, plastics, and other materials that have been left out in space for several months. We want to see how the harsh conditions have changed the materials over time.

2

The flight engineer (see page 8) on board the Space Station has already looked at the samples. But my laboratory here on Earth is better equipped to examine them in more detail. I unpack the samples and inspect the materials.

I specialized in materials science in college before getting my job. I mostly work in a laboratory doing research, conducting experiments, and designing new materials. I'm always learning new things!

3

We want to see if the materials left out in space would be able to protect our spacecraft on long missions. The first test is to see if any of the materials have been worn away. I compare them with photos we took before. Two of the twelve samples fail this test.

4

The second test is to check for damage from fast-moving space particles (tiny objects). Using a microscope, I can see that three samples show signs of damage.

5

I finish the day writing up the report. I just have time to check on another project: a prototype (sample) of a new vest we've designed. It's made from a special material that will protect astronauts from radiation (harmful particles that move through space). I think it looks great!

MY JOB: BEST AND WORST PARTS

BEST: Some of our developments also have important uses on Earth.

WORST: Experiments can take a long time. If one thing goes wrong, I have to start all over again!

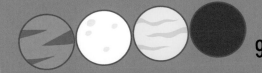

FLIGHT SURGEON

I look after the health of the astronauts before, during, and after their stay on the Space Station. Checking astronauts' health is important. Because the astronauts are weightless in space, they don't have to work as hard to move, so their muscles and bones can become weakened. I help them to stay fit and strong. I work at the space agency—a government organization dedicated to space exploration.

After medical school, I worked as a hospital doctor. However, I've always been fascinated by space so I went back to school for a master's degree in aerospace medicine and, after many years of training, got my job as a flight surgeon.

1

My first job of the day is to meet with my patients on the Space Station. But I'm not up in space! I work at the Mission Control Center on Earth. Every morning, I have a video call with each of the five astronauts to check their health. I talk any worries or problems through with them.

3

Training the astronauts is important since it helps me get to know them really well before they leave. I can't physically examine them when they're up in space, so I need to know as much about them as I can before they go.

2

Next, I work with an astronaut on Earth on her fitness. We design exercise programs for our astronauts for before and during their time on the Space Station. Before this astronaut leaves for space, she needs to work out for several hours a week to maintain her strong muscles and bones.

MY JOB: BEST AND WORST PARTS

BEST: I enjoy helping astronauts to stay fit and healthy during their missions.

WORST: I worry in case there are any emergencies during missions and I won't be able to help.

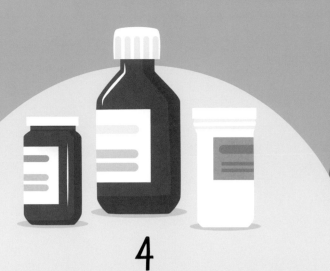

4

After a quick lunch, it's off to a meeting with the medical team. We discuss the medical supplies kept on the Space Station. Space affects astronauts' immune systems: they are less able to fight off illness. A crewless spacecraft with supplies will be sent to the Space Station soon, so we decide to send more antibiotics (medicines that fight infection).

5

My next job is training another crew who will leave for the Space Station in a few months. They need to know what to do in case of a medical emergency. Today, I'm using a dummy to show them what to do if someone stops breathing.

6

Later in the afternoon, I test the fitness of an astronaut who has returned from the Space Station. He had exercised for two hours every day while on the Space Station, and he seems to be in really good shape!

7

It's 5:00 p.m. and time to leave. I pick up a report to read at home. We are always learning more about the effects of space on health, and we now think being in space may affect eyesight. Tomorrow we'll have a meeting to decide how to study this with the next crew.

ASTRONOMER

In my job, I use telescopes to study space. It's fascinating work—can you imagine being someone who explores the universe? Most of the time, I'm based at a university where I teach, study, and analyze data. But, for a few weeks every year, I visit an observatory—a building that houses a huge telescope—that is shared with other astronomers, and that's where I am today.

Space is a big place, so there's a lot to learn! After getting my bachelor's degree in astronomy, it took me another five years to earn a PhD. I then worked as an assistant astronomer until I was more experienced.

1

I start my day at 4:00 p.m. That's right—I wake up in the afternoon! As I eat breakfast, I read up on some new research. As an astronomer, I'm trying to make new discoveries. But I can only do this by building on knowledge from other astronomers, so I need to do a lot of reading.

MY JOB: BEST AND WORST PARTS

BEST: I'm an explorer! I discover things that no one has ever seen before.

WORST: Astronomers take turns using the observatory telescopes, and sometimes I have to wait weeks, or even months, to use one.

2

As the sky begins to get darker, I head to the observatory to start work. Most observatories are in remote places where the sky is clearer, with no pollution or artificial light. It's so beautiful here— I love looking up at the sky on my way to work.

3

My current project is about finding new planets in our galaxy called exoplanets.

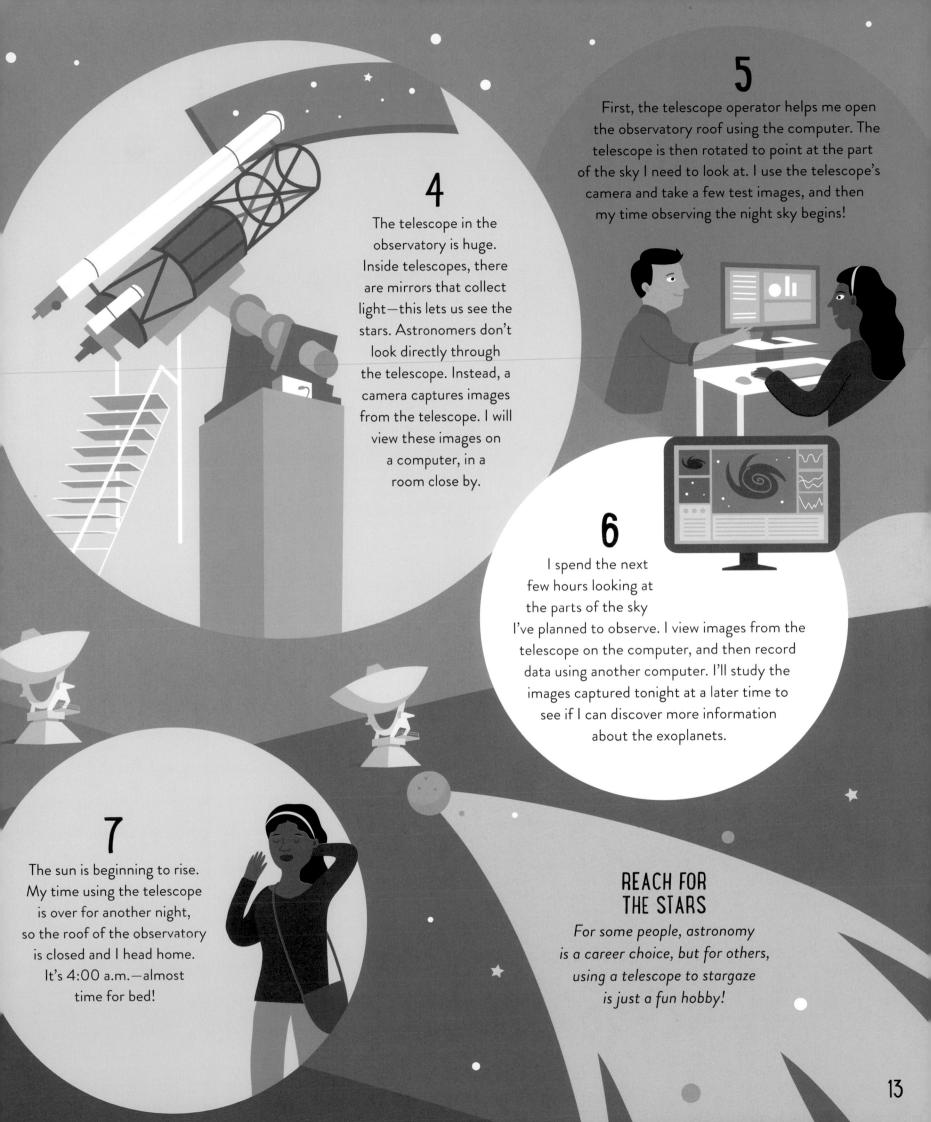

4

The telescope in the observatory is huge. Inside telescopes, there are mirrors that collect light—this lets us see the stars. Astronomers don't look directly through the telescope. Instead, a camera captures images from the telescope. I will view these images on a computer, in a room close by.

5

First, the telescope operator helps me open the observatory roof using the computer. The telescope is then rotated to point at the part of the sky I need to look at. I use the telescope's camera and take a few test images, and then my time observing the night sky begins!

6

I spend the next few hours looking at the parts of the sky I've planned to observe. I view images from the telescope on the computer, and then record data using another computer. I'll study the images captured tonight at a later time to see if I can discover more information about the exoplanets.

7

The sun is beginning to rise. My time using the telescope is over for another night, so the roof of the observatory is closed and I head home. It's 4:00 a.m.—almost time for bed!

REACH FOR THE STARS

For some people, astronomy is a career choice, but for others, using a telescope to stargaze is just a fun hobby!

13

ASTRONAUTICAL ENGINEER

When I was a child, I dreamed of blasting off to Mars. Now, I actually design and build rockets! Some people think you need to be a genius to be a rocket scientist, but you just have to love the job and enjoy learning as much as you can about it.

After earning a degree in astronautical engineering, I joined a space agency internship program that led to me getting my job. A lot of engineers work here and they are trained in different areas, such as robotics, electronics, or computer engineering.

1

It's 9:00 a.m. and I'm at a team meeting, where we all share what we're working on. I am helping to design a new, more powerful rocket that can travel to Mars in half the time our current spacecraft take! It's a really exciting project.

2

My team is working on the rocket's propulsion system, which is the engine—the part that makes it fly! After the meeting, I put on my protective clothing and go to the laboratory to do some tests.

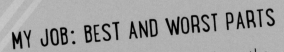

MY JOB: BEST AND WORST PARTS

BEST: Building a machine that *leaves* the planet is the best job *on* the planet!

WORST: Making sure everything works properly can be stressful—there are no repair shops in space.

3

The rocket propulsion system needs to be heated to a very high temperature for it to work properly. I'm testing whether the materials we have designed can cope with that. We have used a strong metal called tungsten and mixed it with other substances.

4

I heat the tungsten to 9,700°F (that's very hot!) and then check it with a microscope. Unfortunately, there's a tiny crack—that means I'll need to think of a way to make the metal stronger.

ANYTHING THAT FLIES

Astronautical engineering is a type of aerospace engineering. Aerospace engineers can also specialize in building airplanes and helicopters.

5

I record the results on a report so I can keep track of what works and what doesn't. We want to learn from our mistakes instead of repeating them!

6

In the afternoon, I discuss the problem of the cracked metal with my team. We decide to slightly increase the amount of tungsten and try again. I'll work on that tomorrow.

7

My final task of the day is attending a training session about a new space probe that has landed on Mars. Space technology is always changing and we have to stay up to date with developments—that's why I find it so exciting!

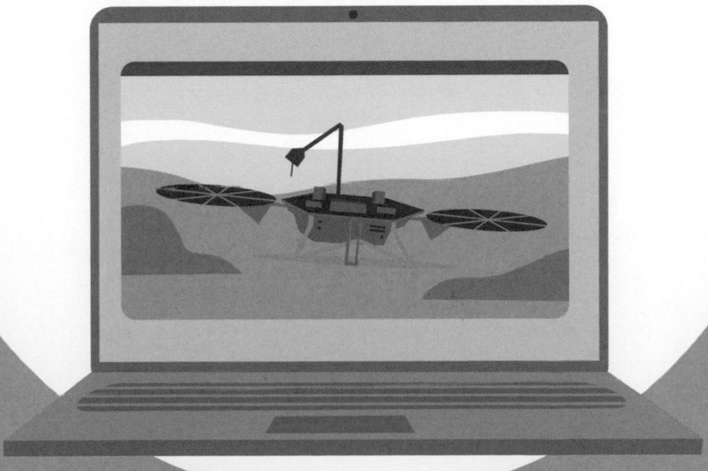

SPACE LAWYER

People are always surprised I'm a space lawyer—most don't even know the job exists. But wherever people go, they need laws and wherever there are laws, there are lawyers ... including in space! After law school, I earned a master's degree in space, cyber, and telecommunications law. Now, I work at a firm with other specialists.

There are a lot of space laws to understand, such as the rescue of stranded astronauts, space exploration, and rules about damage caused by space objects.

1

This morning, I meet with a communications company that wants to send a satellite into space. There are over 2,000 satellites in space, each sending signals down to Earth to connect us to cell phones, the Internet, TV, and radio. I give the company advice on the laws about building and launching the satellite.

2

The company must apply for a license to launch the satellite. I grab my documents and explain how to apply for one. It would be hard for them to understand what to do without my help. Helping people is what I love about my job.

MY JOB: BEST AND WORST PARTS

BEST: I like that I can use my expertise to help people.

WORST: I spend a lot of time thinking about space but never actually going there. Maybe one day I will!

3

Space tourism is fast becoming a reality, so it's an exciting new area for space law. My next meeting is with a new client—a company that wants to send tourists on trips into space!

4

They need me to write a contract that passengers will sign to say they understand the dangers involved in spaceflight. I make notes on what they want to include in the contract, and we agree on a deadline for the work.

5

It's been a busy morning, so I take a quick lunch break with my colleagues. We still talk about space law though, since we all find it so fascinating!

6

Later, I meet my final client of the day—a TV company that wants to rent a satellite to broadcast TV shows around the world. They've asked me to read their contract with the satellite owner to make sure it is fair. I need to pay really close attention—I don't want to miss anything important!

7

I spend the rest of my day doing research into new space laws. As our space technology grows, and as more and more people venture into space, who knows how many new laws we will need in the future?

17

ASTROBIOLOGIST

My job is to investigate the possibility of life beyond Earth. It's a really exciting time to be an astrobiologist—some of the scientists at the space agency where I work are studying the possibility of life on Mars! My job is just as interesting: to look at whether the frozen moons in the outer Solar System could contain life.

After earning a PhD in biology, I went on to specialize in astrobiology. There's a lot to learn—everything from biology, chemistry, and physics, to geology, astronomy, and oceanography.

1

I start my day in the laboratory, checking on an experiment. I'm investigating one of Saturn's moons, called Enceladus. Astrobiologists are interested in how life started on Earth, and we think it may have begun near hot springs under the sea. We believe Enceladus also has these, which means there could be some form of life growing there.

2

For my experiment, I have tried to create similar conditions to the hot springs. I use a powerful microscope to see if chemical reactions might create living things. I don't see any sign of life just yet, so I will need to give it some more time.

3

Next, I meet with a team who are planning a mission to another of Saturn's moons. They show me a model of the probe that will collect the data to help find signs of life. It's very exciting!

4

After lunch, a journalist from a newspaper arrives for an interview. I really enjoy explaining what I do to the public. I tell the journalist that we are looking for tiny living things called microbes, not "space monsters"— even though they can look a bit like that!

5

After the meeting, I head to a conference to give a talk about my research project. I also listen to other astrobiologists talk about their investigations into life on Mars. We all enjoy sharing our research with each other.

6

I finish the day writing a report about Enceladus. It's hard to believe that anything could live on this small moon so far away. However, we know that life exists far below the surface of the ocean on Earth. So why shouldn't it be able to survive in other parts of the Solar System?

SPACE BIOLOGY

While an astrobiologist looks for life that might exist in space, a space biologist studies how actual living things survive in space. For example, biologists are experimenting with growing food in space so that astronauts can have fresh fruit and vegetables.

MY JOB: BEST AND WORST PARTS

BEST: It's like real-life science fiction because I'm searching for extraterrestrial life.

WORST: We haven't found any life yet, so it can be frustrating, but we are always hopeful!

NUTRITIONIST

It's not just flight surgeons who look after astronauts' health. I give advice to astronauts about what to eat when they're in space. Astronauts spend several months on the Space Station, so a lot of planning goes into keeping them healthy and happy for all that time.

I combined my biology degree with my love of food by studying for a PhD in nutrition. I work closely with the astronauts, while other scientists work on developing the food itself, making sure it is safe to eat and that it will last for the entire spaceflight.

1

This morning, I'm in the test kitchen with an astronaut who will be leaving for the Space Station in a few months. She is trying some of the food and drink we have prepared to see what she likes. I will then design a meal plan for her trip. We need to make sure the astronauts actually enjoy what they're eating!

2

A lot of the food tastes good, but doesn't look very nice. This is because we remove the water from it to save space and weight during launch. I show the astronaut how to add water to a package of macaroni and cheese using a special machine. She can then eat it straight from the container.

3

The astronaut selects her three meals and two snacks per day. There's a lot to choose from, including pasta, chicken, fruit, nuts, and even chocolate brownies. Her menu will repeat every eight days so she doesn't get too bored with it!

4

In the lab, I run some tests. I check the astronaut's menu to make sure it has the right vitamins, minerals, and calories to keep her healthy. The tests show that she needs more nutrients.

5

I'll recommend adding more vegetables at our next meeting. The food will be ready a month before launch.

6

I eat lunch, then I check the records of the current Space Station crew. The crew log their meals on an app to show how much they eat and drink every day. I can see that one of the astronauts isn't getting quite enough water, so I'll need to remind him to drink more.

7

Next, I contact one of the astronauts on the Space Station. He's losing too much weight. When the next supply spacecraft is sent to the Space Station, I'll include some foods with more carbohydrates and fat to add to his diet to help him put on weight.

8

Before I go home, I check on some test results in the laboratory. In space, astronauts suffer from bone loss, when the bones weaken. We put two astronauts from a previous mission on a special diet, which included more fish. The results of the test show that they suffered less bone loss. This is good news for future astronauts … except for those who don't like fish!

MY JOB: BEST AND WORST PARTS

BEST: It's exciting to find new ways to keep our astronauts healthier and happier in space.

WORST: I wish the Space Station had a real kitchen so we could create fresher food. Let's hope that can happen in the future!

FLIGHT CONTROLLER

I work in a Mission Control Center on Earth. My job is to keep astronauts safe on the Space Station and deal with any problems. I work as part of a team of flight controllers, and we do a variety of important jobs, from looking after the Space Station's power to monitoring its air systems so the astronauts have air to breathe. It's an amazing career!

I studied engineering in college, then did lots of training exercises at the space agency to qualify as a flight controller. For example, I practiced what to do if the equipment went wrong or if there was an emergency such as a fire on board.

1

I start the day working in the mission control room, where we each have our own console (computer desk). We get an incredible view of Earth on the big screen. I can even make out a hurricane spinning across the ocean!

2

I'm in charge of the Space Station's communication systems. We work in shifts, as there needs to be someone keeping an eye on things all the time! My first job is to speak to the controller I'm taking over from to see if there have been any problems. Luckily, all seems to be well.

3

Next, I read through the log. This is a record of everything that happens during a mission. If something goes wrong, we'll be able to check the log to see what the problem was. I'll continue updating the log during my nine-hour shift.

4

I sit in front of the screens and monitor the data. Everything works well for the first part of my shift. Then suddenly, I see an alert signal on my computer. One of the Space Station's communication antennae isn't working properly. Fortunately, we always have a backup, but the antenna will still need to be fixed.

5

I speak to the flight director, the most senior flight controller who leads our team, and she calls us together so we can decide what to do. Communication skills are vital in this job. We have to work as a team when things go wrong!

6

After the meeting, I call the astronauts and explain the situation. I have to remind myself that I'm talking to people who are hundreds of miles above us, orbiting (moving around) Earth at over 17,000 miles per hour!

7

We need to examine the broken antenna and take pictures so that we can figure out how to repair it. Of course, this is not a simple job; it's out in space! We decide that one of the astronauts will need to take a space walk. I explain to him where the antenna is and what he needs to do.

MY JOB: BEST AND WORST PARTS

BEST: I love being part of such a brilliant team, working and solving problems together.

WORST: It can be stressful when something goes wrong so far away. But we usually work out a way to fix it!

8

It takes 24 hours to prepare for a space walk. I write up all the details in the log. Tomorrow, I'll be guiding the astronaut on his space walk. After a long shift, it's finally time to hand over to another controller. After telling him what has happened, I leave for home, excited about the challenge facing me tomorrow.

ASTROPHYSICIST

How and when did the universe start, and how did it make all those galaxies, stars, and planets? Ever since I was a child, I've wanted to know the answers to these questions. As an astrophysicist, it's my job to spend most of my time figuring them out!

I've always had a passion for math and physics, as well as for space. I studied astronomy in college, then earned a PhD in astrophysics. It took many years of hard work and study to reach my position as a university research scientist, but I've enjoyed every minute!

1

I arrive at my office and sit in front of the computer. I spend more time looking at computers than through telescopes! The universe is getting bigger all the time, and I study data to understand how this is happening.

2

My area of interest at the moment is quasars. These are bright objects, a billion times bigger than the Sun. The Hubble Telescope, a huge telescope that orbits the Earth, gives us lots of information about quasars.

3

I study the most recent data sent from the Hubble. The light from the quasars has taken billions of years to reach the telescope. I spend the morning calculating these distances to help me to understand how fast the quasars are moving away.

MY JOB: BEST AND WORST PARTS

BEST: My job is all about asking questions nobody knows the answers to, so it's a real adventure.

WORST: Because I'm trying to understand new information, it can get a bit mind-boggling!

4

I take time out from research to record a science podcast. I spend half an hour talking about everything from the planets to black holes. It's a really fun part of the job—I love sharing fascinating facts about space.

5

It's time for lunch. I meet with my colleagues in the cafeteria. We discuss a report I'm preparing for an important science publication. Astrophysics is a huge subject, so we need to work together to share our ideas.

6

In astrophysics, there's always more to learn. After lunch, I walk over to a lecture hall to attend a talk given by a leading astrophysicist about black holes. I find it fascinating and I take lots of notes!

7

As well as doing my own research, I also supervise a group of junior researchers. I meet with one of them to talk about her work. She wrote an interesting report about different kinds of galaxies. She's always asking questions and is really keen to learn.

8

Finally, it's time to go home. I wonder what to have for dinner and decide on pizza. That was a slightly easier question to answer than "How did the universe start?"

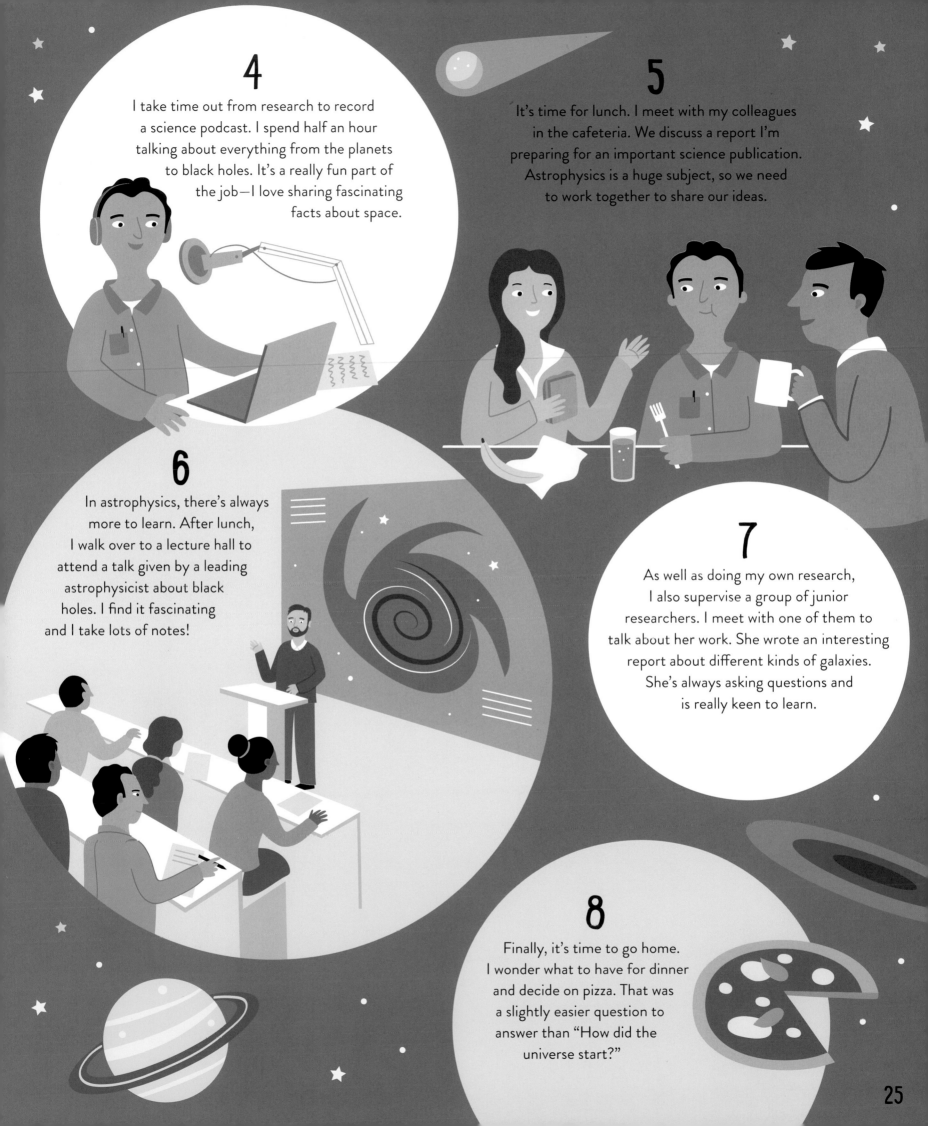

SPACE CENTER MANAGER

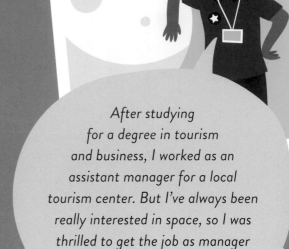

People come to our space center to learn all about space, from astronauts and rockets to planets and stars. My job is to manage the exhibitions, supervise the other staff, and make sure the visitors are happy. It combines tourism, business, and science, so I have to be good at working with people, dealing with money, and know lots about space!

After studying for a degree in tourism and business, I worked as an assistant manager for a local tourism center. But I've always been really interested in space, so I was thrilled to get the job as manager here at the space center!

1

We have hundreds of visitors each day, so every morning, I walk around the exhibitions to make sure everything is ready to go. I greet my staff and ask if they have any questions. As a manager, I need to make sure the center is a happy, friendly place to work!

2

At 10:00 a.m., I meet with the exhibition organizer. Our national space agency is launching a solar probe (a spacecraft that will explore the Sun), so we're going to have a big display all about the Sun. We talk about ideas for interactive activities, such as a model spacecraft that uses solar power.

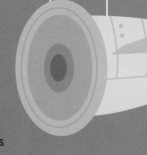

3

After the meeting, I visit our gift store to talk to the store manager about how sales are going. She tells me that the science kits are selling really well, so we agree to add more to the next order. Along with keeping the customers happy, we need to bring in as much money as possible to support our center.

4

In the afternoon, I join the marketing officer for a visit from the local TV station. Our latest exhibit is amazing; it's a spacecraft capsule. It was never launched into space because the mission was abandoned, so we bought it. We are arranging lots of media interviews to tell people about it, in the hope that they'll come and enjoy our space center.

5

My day finishes with talking to some visitors about the latest space explorations. I really enjoy passing on my space knowledge to our visitors.

SCIENCE LAB

MY JOB: BEST AND WORST PARTS

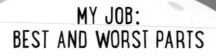

BEST: I love being around other people, and the space center is always lively with visitors!

WORST: I get asked some questions about space I don't always know the answers to!

PLANETARIUM EDUCATOR

People come to a planetarium to learn more about the wonders of our universe. I run shows and events for visitors of all ages. The building has a domed roof to show the night sky. I love talking to the audience and watching their amazement. It's a magical experience!

I earned a degree in physics, then did further study to earn an advanced degree in astronomy. I also worked as a volunteer tour guide at a local museum, so I knew I wanted to work with people.

1

My day begins by making sure everything is set up and ready for today's show. A group of excited schoolchildren arrive. They settle into their seats as I turn off the lights. It's wonderful to hear the children's gasps of amazement as the night sky appears on the ceiling above them.

2

I tell them about stars in the sky and show them how to find Polaris, the North Star. It always looks like it's in the same place above Earth, so it can be used to figure out which way is north.

3

When the show is finished, I take the children on a tour of the planetarium to show them some of the exhibits. They particularly like looking at the giant telescope!

MY JOB: BEST AND WORST PARTS

BEST: There are always new discoveries in astronomy, so I'm a student as well as a teacher.

WORST: It can be frustrating when bad weather ruins an outdoor event I've planned.

4

After lunch, I take time to answer emails and work on our monthly blog, which tells visitors what to look for in the night sky. I also do some research, to make sure I'm up to date with the latest discoveries and theories in astronomy.

5

My last job of the day is to meet a group of adults for some real-life stargazing outside. Luckily, it's a clear evening, so we get an amazing view of the night sky!

27

SPACE WEATHER FORECASTER

I've always been interested in space, but I didn't plan on becoming a space weather forecaster. After studying physics in college, I joined the government's weather forecasting service. That's where I found out about the space weather department. I was so fascinated that I started working here.

The weather on Earth is affected by what happens in space—especially by the activity of the Sun. It's important to monitor space weather so that we can be prepared for what it may bring, such as huge storms.

1

I arrive for my shift early in the morning. My colleague has worked through the night, and now she can go home! We monitor the weather 24 hours a day, so we take turns to work at night.

2

My first job is to look at data gathered by the weather satellites. These monitor the weather in space and send us lots of information, such as how fast the solar wind is traveling. This wind travels toward Earth at huge speeds, and can cause damage to spacecraft. There is nothing unusual today, with the wind traveling at a normal 900,000 miles per hour!

3

After I have collected the information, I send it to other forecasters in different space weather centers across the world. This job is all about the whole planet working together.

MY JOB: BEST AND WORST PARTS

BEST: There is still so much more to discover about space weather—and that's why I find it so exciting!

WORST: Not many people know about my job, so I often find myself explaining that there is actually weather in space!

4

After lunch, I take a look at some new images of the Sun. I check the surface for sunspots. These darker patches on the Sun's surface can help us to predict the Sun's activity. Where there are sunspots, there may be solar flares.

5

Solar flares are bursts of energy from the Sun. They can cause damage to satellites in space and communication systems (such as radio signals) on Earth. These bursts can also harm the health of astronauts, so we need to look out for them. I can see in the images that there has been a solar flare recently, so I make a note to track its impact and discuss with colleagues.

6

After looking at the data and images, I type up the space weather report, which will be posted on our website.

7

Just before going home, I have a meeting with a team of scientists. We talk about a new instrument we are developing that will give us early warnings of any disruptive space weather on its way. We are always working on new equipment and techniques to help us do our job as best we can!

WORLDS OF WEATHER

Scientists also study the weather on other planets. For example, there is a huge storm on Jupiter that has lasted for over 400 years, and winds on Saturn can reach 1,000 miles per hour.

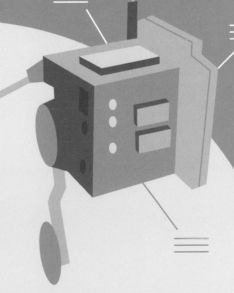

PUBLIC AFFAIRS OFFICER

I'm the main point of contact between the space agency and the public. I manage our website, answer emails, make videos, and find lots of other ways to let people know about all the exciting things happening in space exploration. What we do at the agency should be open to everyone to learn about—it's my job to make that happen.

I studied media production in college and learned all about film, photography, website design, and different ways to communicate with the public. In my job, you need to enjoy working with people and be really organized, since there are always so many different things to do!

1

I'm part of the team that manages the space agency's education website. This morning, I'm meeting an astronomer working with one of our largest space telescopes. She chooses some photographs for the website.

2

After the meeting, I post the photographs and write an explanation to help students understand what the pictures are showing.

3

Next, I check some competition entries. We asked school students to name a rocket carrying a space probe (robot explorer) headed for the Sun. There have been thousands of entries! We've picked our favorite—"New Dawn"—and I call the winner to let them know that their prize is to attend the launch of the rocket. Of course, the winner is delighted!

4

Next, I take a call from a local museum. They want to borrow some old space suits for an exhibition. Believe it or not, we do lend these out. I email them an application form and also ask whether they want us to provide a speaker.

5

After a quick lunch, it's off to meet one of our filmmakers. We are making a video about sending humans to Mars. I interview a scientist with questions sent in to us by the public.

6

The space agency has its own TV channel, so my next job is to visit the TV manager's office. A show about our first all-female space walk is coming soon—exciting! We talk about the show, and I make some notes for a blog post.

7

It's the end of the day and I finish by updating our social media. Our followers love hearing about what to look for in the night sky. I write a post to let them know that Mars, Jupiter, and Saturn will be visible in the sky just before dawn tomorrow. I'll be up early to have a look, too!

MY JOB: BEST AND WORST PARTS

BEST: I really like creating the educational materials that help children learn about our universe.

WORST: Sometimes I can get a bit nervous before I have to talk on camera!

SCIENTIFIC RESEARCHER

I'm part of a science research team that specializes in studying Mars. Right now, I'm on a research trip in Antarctica. Antarctica is very cold and dry, which makes it perfect for learning about extreme environments in space, and testing equipment we might take there. I've always been interested in space, so I love my job as a researcher—even when I'm freezing!

1

Like most people, I start the day by getting dressed for work. This takes quite a long time. First, I put on thermal underwear. Next, I wear a warm sweater and pants, then my thick, waterproof jacket and pants. Finally, I pull on two pairs of gloves and my knit cap. Now I'm ready to go!

With my degree in astronomy and further specialist degree in planetary science, I was able to get my job at the space agency research center. All the scientists that work there are passionate about learning and finding out as much as possible about the universe.

2

After a short trip on my quad bike to a huge, rocky valley, I stop and take out a drill. On Mars, we plan to drill into the ground to examine the rocks and search for signs of life. I'm testing the long drills designed to do this.

3

I assemble the tool and it drills down about 13 feet. It scoops out samples of rock and soil, and empties them into a container. The drill seems to work really well.

4

Back at the research center, I carefully store the samples. I'll send these to a college that is interested in using them to find out what kind of life exists in the Antarctic. I like the fact that my research is also helping us learn more about our own planet.

5

Instead of a tent, I'm staying in an inflatable base that has been designed to be used by astronauts on the moon. It's been designed by colleagues on my team, and there have been a few versions of it. I'm testing the latest version on this trip—the moon, like Antarctica, is a very cold and dry environment, so my colleagues are keen to hear how it goes!

6

After a long day, I settle down for the evening. It's chilly out there, but my tent and sleeping bag keep me protected and warm. I drift off, happy to know that even by just sleeping I am doing important research!

MY JOB: BEST AND WORST PARTS

BEST: I'm helping the human race to prepare for exciting future missions.

WORST: A lot of research jobs are temporary, so I often have to move to a new job when a project is completed.

ON EARTH AND IN SPACE

In addition to missions on Earth like this one, scientific researchers also analyze data sent back from exploration spacecraft to help understand the atmosphere, weather, and surface of Mars.

SCIENCE FICTION AUTHOR

Space exploration, time travel, aliens on other planets ... I've always loved reading science fiction! But I've also always been interested in real space and science. That's good because I spend a lot of time doing research to get all the details right. Of course, I've always loved writing too!

You don't need specific qualifications to be a writer—but it does take skill and hard work. Getting your writing noticed by book publishers can take a long time. In this job, you need a lot of determination and self-belief.

1

I'm writing a new book about a journey to Saturn. This morning, I read what I wrote yesterday. It's amazing how many changes I need to make!

2

Mid-morning, I take a quick break. I find that working in short bursts is better for me, so I stop for a snack. I need a lot of self-discipline; after all, I don't have a boss to make me write, or to check what I'm doing.

3

I begin my research for the next chapter, where I'll write about Saturn's atmosphere. I need to make sure all my facts are correct, so I write a list of questions and email them to a scientist at a local college.

4

I write into the afternoon, eating lunch while I work. I love using my creativity to imagine new worlds, so the time flies by.

5

When I finish, I head to my local bookstore. I've been invited to sign copies of my latest novel for customers. After spending all day in front of my computer, it's great to get out and meet my readers!

MY JOB: BEST AND WORST PARTS

BEST: It's really exciting when I see one of my books out in the world!

WORST: I work by myself, so some days I don't speak to anyone all day.

TECHNICAL WRITER

I've always liked science as much as writing. When I saw that the space agency was looking for a technical writer, I realized it was a great way to combine my interests. In my job, I make complicated information easier and simpler for people to understand. If you enjoy a challenge, you'll like doing this job!

1

We are sending scientific equipment to the Space Station, so I'm preparing the instruction manuals for the astronauts. The instructions from the developers who created the equipment are very long and complicated, so I've been working to make them shorter and easier to understand.

I earned a degree in English, followed by a master's degree in technical writing. I work with a variety of different people at the agency, such as scientists, designers, and engineers. You need to be able to work as part of a team in this job.

2

After completing the first section of the documents, I send it to the developer for checking. Next, I have a training session with another developer about some of the newest equipment. I need to properly understand how it works before I write more.

3

I spend the rest of the afternoon writing up the new instructions, checking my work, then checking it again and making changes. Attention to detail is really important!

4

My last meeting of the day is with the graphic designer. The manuals will need diagrams and charts, which the designer will be creating. I enjoy our meetings and we've worked together for a long time—the space agency needs a lot of instruction manuals!

MY JOB: BEST AND WORST PARTS

BEST: It's a really interesting challenge to help people understand complicated information.

WORST: I'm always racing against the clock to finish my work by set deadlines.

SPACE SUIT DESIGNER

After studying mechanical engineering in college, I never thought I'd end up designing clothes! But space suits aren't like regular clothes. They are specially made to protect astronauts from extreme conditions, such as freezing temperatures, while allowing them to move around. It's a real engineering challenge!

1

I arrive at the space center laboratory and continue working on a new suit that will be used for walking on the moon's surface. The suits we currently use were designed for space walks, where astronauts only go outside the Space Station, and they aren't very easy to move around in. Today, I'm looking at the new suit's glove design.

After college, I joined an education program at the space agency, then got a job working with the space suit design team. I really enjoy designing the world's most expensive clothes!

2

Gloves are made from layers of tough fabric but need to be flexible enough for astronauts to handle tools and operate controls. I spend the morning working on some designs using a special glove box that copies the conditions in space.

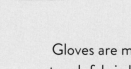

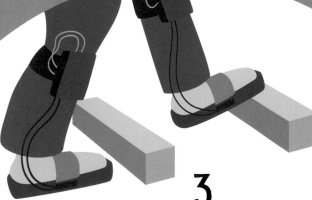

3

After lunch, I test space boots! I'm excited because it's taken a long time to reach this stage. I put on the boots and find that they're quite easy to move around in, just as we designed. To help stop stumbles, the boots have sensors that vibrate when the astronaut walks toward a rock or other obstacle. The sensors seem to be working fine, too.

4

I finish the day planning the next part of our space suit—the built-in bathroom! Astronauts may spend hours outside their spacecraft, so we need to work on ways they can "go" while wearing their suit. It's amazing the things we have to think about in this job!

MY JOB: BEST AND WORST PARTS

BEST: It's the best feeling when I see astronauts actually wearing my creations in space.

WORST: Space suit design takes years; this isn't for those who like fast-paced fashion design!

COMPUTER ENGINEER

Everything in space depends upon computers, from space suits to spacecraft to communications. I design computer hardware, the actual computer equipment. Other specialists design software, the programs the computers run. I'm working on a new computer monitoring system that will measure spacecraft air quality, to make sure the astronauts stay safe and healthy in space.

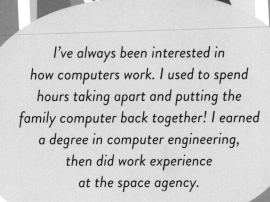

I've always been interested in how computers work. I used to spend hours taking apart and putting the family computer back together! I earned a degree in computer engineering, then did work experience at the space agency.

1

The day starts with a meeting about the new computer equipment with scientists and other engineers. It will take a long time to develop the new hardware, so there will be lots more meetings over the coming months until everything works perfectly. Luckily, the team gets along!

2

After the meeting, I study the system we currently use to monitor the air quality in order to figure out what works well and what needs to improve. Everything computer engineers do is about trying to make things work better.

3

I have lunch, then head to the gym. I spend a lot of time in front of a computer or looking closely at hardware, so I try to be active when I can.

4

In the afternoon, I meet the software engineer to discuss my first ideas for the new monitoring system. Computers are useless without instructions—the software—so we'll be working very closely together.

5

Before I head home, I speak with a company that has designed a new microchip. This is a tiny piece of electronic equipment that fits inside computers to pass on information. The new microchip is made of a material that will make our computers work much faster. Computer science is a fast-moving field, so I have to be able to keep up with the latest developments. That's what's exciting about it!

MY JOB: BEST AND WORST PARTS

BEST: Since everything relies on computers, I'm at the heart of the mission.

WORST: Building a computer is quite straightforward, but it can take a long time to get it working properly!

ASTRONAUT INSTRUCTOR

I've loved scuba diving since I was a teenager. After studying mathematical sciences in college, I applied as a diving instructor for the space agency's astronaut training facility. Here, we prepare astronauts for what it's like to be weightless in space—by using a huge swimming pool!

As well as diving instructors, there are other kinds of astronaut instructors, such as those who train astronauts on how to use their space suits, how to use the spacecraft systems, or how to carry out experiments in space.

1

We start by checking that all our training equipment is ready and safe for the astronauts to use. Safety always comes first, whether in space or underwater. The training takes place on a model of part of the Space Station, which we placed in the water earlier.

2

I have already trained the astronauts in scuba diving, so they are now ready to practice a space walk— moving around the outside of the Space Station. My team and I are assigned an astronaut to look after. Once underwater, we adjust the weights on her suit so that she feels completely weightless, just like in space.

3

The astronaut practices making repairs on the outside of the spacecraft. I stay near her all the time, checking on her safety.

4

After the session, I join my co-workers for a rescue exercise. A space suit has sunk to the bottom of the pool, and we have to bring it up quickly. It's really heavy, but after all our training, we're strong enough to carry the suit up to the water's surface.

5

After several hours in the water, I'm ready to hang up my scuba gear, jump in the shower, then head home for a good night's rest!

MY JOB: BEST AND WORST PARTS

BEST: I help astronauts be the best they can at their job.

WORST: I'll never get to experience actual "space diving."

NEUROSCIENTIST

As a neuroscientist, my job is to study how astronauts' brains and nervous systems are affected by the extreme environment of space. The data I collect will help the space agency figure out ways to improve the astronauts' physical and mental health. Like space, the ocean is an extreme environment for humans to live in, so being beneath the waves is the best way to test!

I got my PhD in neuroscience because I'm fascinated by people and how they think. I work with astronauts before and after missions. I'm also an experienced scuba diver, which is how I ended up doing my current job!

1

I'm about to enter our research base at the bottom of the ocean. The living quarters here are small and cramped, and we're a long way from civilization—it's similar to living in space. I'll be recording any changes in how the astronauts think and feel about being here, and how they work in an isolated place. We'll be here for three weeks.

2

Once we're settled in, I prepare our first experiment. The astronauts are given special tools to collect samples from the coral for research. I watch to see how working in this environment affects their ability to perform these tasks.

3

The astronauts use a waterproof tablet to record how easy or difficult the tasks were. We then move on to some more tests.

4

After the tests are finished, we change out of our diving gear to eat dinner (freeze-dried food with added water—just like in space!). I check in with the astronauts to see how they're doing, and we talk through any difficulties they've had during the day.

5

Later that evening, I write up my notes. I am also studying how sleep is affected by living here. Tonight, we are all exhausted and fall asleep right away, forgetting we are even underwater!

MY JOB: BEST AND WORST PARTS

BEST: It feels like a real adventure—all in the name of science!

WORST: Sometimes, it's hard being away from my family.

EQUIPMENT SPECIALIST

There is so much complex equipment used in space that it needs specialists who really understand it—like me! After I earned my degree in engineering, I started working for the space agency on life-support systems: providing astronauts with the air (oxygen) and water they need.

1

This morning, I'm testing a new breathing system to be used on the Space Station. The system changes the gas carbon dioxide—breathed out by the astronauts—into oxygen for them to breathe in. The test shows that half the carbon dioxide has changed back into oxygen. This is a great result!

Equipment specialists become experts in one area, such as space suits, rocket launch equipment, communication systems, or robotics. Whatever field they work in, they need to be really interested in engineering and design.

2

I didn't create this equipment on my own—a team of scientists and engineers from around the world helped, too. We set up a video meeting so I can show them the results and talk about ways to improve the system. The new machine is too big, so we discuss how to make it smaller.

3

After lunch, I hold a training session to show the astronauts how the machine works. It's important that they understand the equipment, in case something goes wrong in space.

4

In the afternoon, I continue with a new project I'm working on: improving the water recycling system on the Space Station. At the moment, we use special machines to recycle most of the water on board. However, if humans are going to travel as far as Mars, we will need a better recycling system, since it won't be possible to send extra water that far.

5

After a busy day, I head home to eat dinner with my family ... and fix the TV. I'm an equipment specialist at home, too!

MY JOB: BEST AND WORST PARTS

BEST: It's taken a lot of hard work to get here, and now I really enjoy being a leading expert in my chosen field.

WORST: Working on machines that keep astronauts alive can be a lot of pressure—I can't afford to make any mistakes.

TECHNICIAN

I've always been fascinated by spaceflight—that's why I love working at the wind tunnel research center. Here, we examine how vehicles behave when they're traveling through air. As a technician, I support engineers and scientists with their work by setting up and monitoring experiments, and looking after the equipment.

I worked hard in math and science at school, then earned my degree in electrical engineering. I was lucky enough to get an apprenticeship with the space agency, which led to my job as a technician.

1

Today, we're testing a model of a new space launch rocket. This type of powerful rocket will be sent to the moon on the next mission, and maybe even to Mars one day! But first, we need to run lots of tests to make sure that it will fly correctly and safely.

2

I fasten the model in place in the wind tunnel—a huge tube with moving air inside it. It needs to be secure, so it doesn't blow away. It's much easier to test a model that stays still, rather than one moving at supersonic speeds! I also tie threads to the model so we can see how the air moves around it.

3

I start the huge fans at the end of the tunnel. These can blow winds of up to 4,000 miles per hour! The air around the model shows what would happen if it was actually moving through the air.

4

The threads aren't showing the air flow well enough, so the scientist leading the test asks me to add smoke so they can more clearly see how the wind flows around the rocket.

5

The rocket is shaking a little in the wind. The scientist wants to run more tests tomorrow to find a way to stop it from happening. We need to make completely sure that the rocket will fly perfectly through the air, to keep the astronauts safe on their next mission.

MY JOB: BEST AND WORST PARTS

BEST: I really enjoy working with a team of scientists and engineers.

WORST: There's a lot of stopping and starting—it takes time to get it right.

PROJECT MANAGER

As a project manager, I help to produce Mars exploration vehicles, called rovers. Sending a rover to Mars is expensive and complicated, so everything has to be very carefully planned—and that's where I come in! I decide who is doing what, keep track of progress, deal with problems, and make sure the project will be ready for the launch date.

1

Our Mars rover is about to have its first test-drive. I meet with the chief engineer, who was in charge of designing the rover. We've been working on it for months, and she's feeling nervous. I reassure her and tell her that I think it will be a good day. It's my job to keep my team's spirits up—even if I am feeling a little nervous myself!

After studying engineering and doing work experience, I was lucky enough to get a job at the space agency. Over the years, I worked my way up to project manager. My passion for planning and organizing means it's the perfect job for me.

3

We watch as the navigation team controls the vehicle. It moves forward, turns left and right, and goes up and down ramps. The control systems are working well.

2

I head to the laboratory viewing room to watch the test-drive with my team of scientists and designers. I'll watch from here and will look for any problems that might come up during the test.

4

The rover has a robotic arm with a camera that lets scientists on Earth see where it is going. We run lots of tests to check that the arm moves smoothly. It's working well, but it's moving a little slower than it should be, so I make a note to speak to the team. If there's a problem, it's my job to plan how the team should work together to fix it.

5

At the end of testing, I meet with the engineers for a full report. If we fix a few small problems, we'll meet our deadline. I'll talk about the issues with the engineers again tomorrow, and will schedule tasks for everyone on the team. I'm so relieved that there are no major problems—the vehicle is nearly ready for launch!

6

I'm so proud of my team. They've worked hard, and it's paid off! We head out to celebrate and grab a bite to eat. Managing isn't just about getting results—it's about making sure your team feels happy at work while they get the job done, and it's a part of my job that I love.

7

Tomorrow I'll begin to organize the transfer of the rover to the launch site. The rover will land on Mars seven months later. Then, who knows what we might discover there?

MY JOB: BEST AND WORST PARTS

BEST: I enjoy leading huge, complicated projects that help us discover more about space.

WORST: With space projects, there are often extremely complex problems to solve, so it can be very stressful to meet deadlines.

YOUR PERFECT JOB MATCH

With so many different jobs to think about, it can be tricky to choose the right one for you. This guide will help you match your skills, interests, and personal qualities to see which job might suit you.

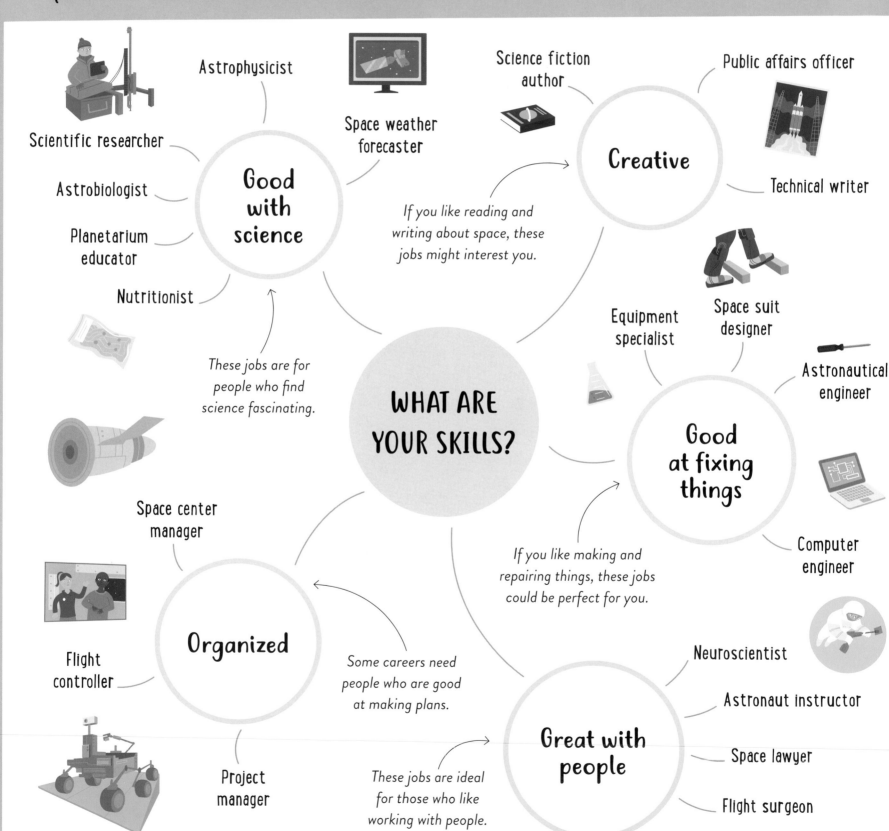

Astrophysicist

Scientific researcher

Astrobiologist

Planetarium educator

Nutritionist

Good with science

These jobs are for people who find science fascinating.

Space weather forecaster

Science fiction author

Creative

If you like reading and writing about space, these jobs might interest you.

Public affairs officer

Technical writer

Equipment specialist

Space suit designer

Astronautical engineer

Good at fixing things

If you like making and repairing things, these jobs could be perfect for you.

Computer engineer

WHAT ARE YOUR SKILLS?

Space center manager

Organized

Flight controller

Project manager

Some careers need people who are good at making plans.

Neuroscientist

Astronaut instructor

Great with people

Space lawyer

Flight surgeon

These jobs are ideal for those who like working with people.

Caring

Nutritionist

Flight surgeon

Flight controller

Planetarium educator

Neuroscientist

Astronaut (flight engineer)

These jobs may be for you if you like taking care of people.

Team player

Astronaut (commander)

Space center manager

Consider these jobs if you like team sports and working with others.

WHAT QUALITIES DO YOU HAVE?

Project manager

Astronaut instructor

Patient

Computer engineer

Technician

Taking the time to get things right is important in these jobs.

Science fiction author

Independent

These jobs are perfect if you prefer working on your own.

Technical writer

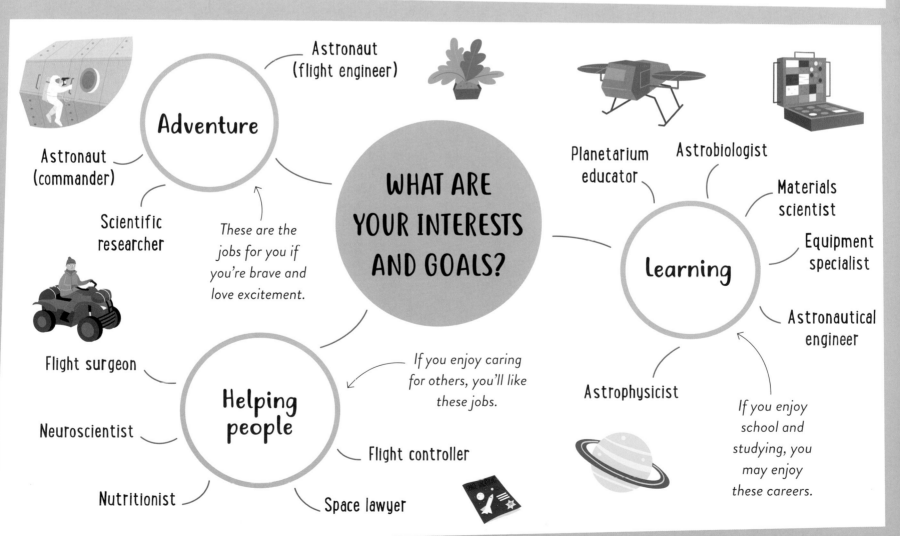

Astronaut (flight engineer)

Adventure

Astronaut (commander)

Scientific researcher

These are the jobs for you if you're brave and love excitement.

Flight surgeon

Neuroscientist

Helping people

Nutritionist

WHAT ARE YOUR INTERESTS AND GOALS?

If you enjoy caring for others, you'll like these jobs.

Flight controller

Space lawyer

Planetarium educator

Astrobiologist

Materials scientist

Equipment specialist

Learning

Astronautical engineer

Astrophysicist

If you enjoy school and studying, you may enjoy these careers.

THERE'S MORE ...

You've read about a lot of different space careers in this book, but there are even more to choose from—and there will be others in the future that we haven't even thought of yet! Read on to discover seven more exciting jobs for space lovers.

ROBOTICS ENGINEER

Many space expeditions are operated by robots. Robotic engineers design and operate the machines that help us to explore space. They might steer vehicles across Mars from Mission Control, or design robotic arms for fixing equipment up in space. They work with very complex systems, so they need to study really hard to be qualified for this job.

JOURNALIST

For those who love writing and have a passion for space, this can be a dream job! Journalists specializing in space write articles and interview people for newspapers, magazines, and websites. They need to be interested in science, since they often write about new scientific discoveries and equipment. Some journalists also write books about space.

METEORITE HUNTER

Sometimes, small rocks called meteorites fall from space to Earth. Meteorite hunters use magnets or metal detectors to track down pieces of these rocks, then they sell them to collectors and museums. This is definitely a job for someone who loves travel and adventure—and rocks!

SPACE ARCHITECT

Space architects work on everything from launchpad design to developing bases that can be used by astronauts on the moon or Mars on future missions. Space architects combine their building design skills with an understanding of the space environment and how it affects humans.

SATELLITE ENGINEER

There are thousands of satellites in space and, with hundreds more being launched every year, designing them is very important. Satellite engineers create satellites for things such as radio communications, ship navigation, and weather forecasting. In addition to engineering, they also need to know coding and math.

CYBERSECURITY SPECIALIST

A hacker is someone who uses coding to get unauthorized access to computer systems. If someone hacked into a satellite, they might disrupt the systems on a spacecraft. Cybersecurity specialists use their computer skills to build defense systems to beat the hackers and protect satellites.

ASTROPHOTOGRAPHER

Using specialized photography equipment, astrophotographers capture images of the stars. The photographs taken can be used by educational organizations as research, or sold as pieces of art. Astrophotographers need good research skills, as well as lots of photography experience. They also need to be patient, as it can take a long time to get the perfect picture!

47

First American Edition 2021
Kane Miller, A Division of EDC Publishing

Copyright © 2021 Quarto Publishing plc

For information contact:
Kane Miller, A Division of EDC Publishing
PO Box 470663
Tulsa, OK 74147-0663
www.kanemiller.com
www.edcpub.com
www.usbornebooksandmore.com

Library of Congress Control Number: 2020936375

Manufactured in Guangdong, China CC 102020

ISBN: 978-1-68464-167-3

1 2 3 4 5 6 7 8 9 10

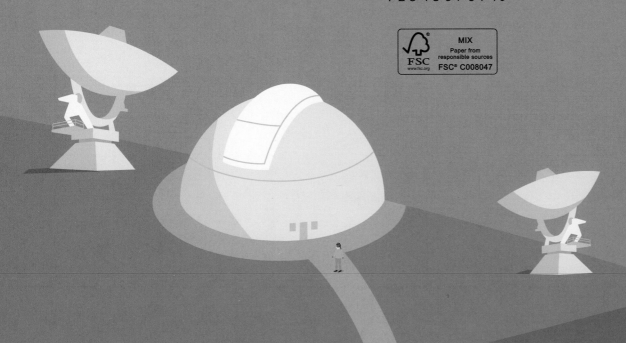